# Three Rich Sheep

written by Lisa Shulman
illustrated by Ellen Joy Sasaki

**SAXON**
PUBLISHERS

# THIS BOOK IS THE PROPERTY OF:

| STATE_____ | Book No. _____ |
| PROVINCE_____ | Enter information |
| COUNTY_____ | in spaces |
| PARISH_____ | to the left as |
| SCHOOL DISTRICT_____ | instructed |
| OTHER_____ | |

| ISSUED TO | Year Used | CONDITION ||
|---|---|---|---|
| | | ISSUED | RETURNED |
| ............... | ....... | | |
| ............... | ....... | | |
| ............... | ....... | | |
| ............... | ....... | | |
| ............... | ....... | | |
| ............... | ....... | | |
| ............... | ....... | | |
| ............... | ....... | | |
| ............... | ....... | | |

**PUPILS to whom this textbook is issued must not write on any page or mark any part of it in any way, consumable textbooks excepted.**

1. Teachers should see that the pupil's name is clearly written in ink in the spaces above in every book issued.
2. The following terms should be used in recording the condition of the book: New; Good; Fair; Poor; Bad.

Three sheep chat on the green grass.

"See the map!" chant the sheep.

The sheep look at the map.
"We can be rich!" chant
the sheep.

Three sheep go on a ship.
They take the map
with them.

3

The sheep see three trees.
"This is where we must
stop!" chant the sheep.

The sheep check the map.
They need to dig here.

The sheep dig a deep hole.
They dig up seeds and
green weeds.

Where is the chest?
At last they see it.

Look at the three rich sheep!

**The End**

## Understanding the Story

*Questions are to be read aloud by a teacher or parent.*

1. Who finds the map?
2. Where do the sheep go?
3. What do they find there?

Answers: 1. three sheep 2. to an island 3. a treasure chest

---

**Saxon Publishers, Inc.**
Editorial: Barbara Place, Julie Webster, Grey Allman, Elisha Mayer
Production: Angela Johnson, Carrie Brown, Cristi Henderson
**Brown Publishing Network, Inc.**
Editorial: Marie Brown, Gale Clifford, Maryann Dobeck
Art/Design: Trelawney Goodell, Camille Venti, Joan Asikainen
Production: Joseph Hinckley

© Saxon Publishers, Inc., and Lorna Simmons

All rights reserved. No part of the material protected by this copyright may be reproduced or utilized in any form or by any means, in whole or in part, without permission in writing from the copyright owner. Requests for permission should be mailed to: Copyright Permissions, Harcourt Achieve Inc., P.O. Box 27010, Austin, Texas 78755.

Published by Harcourt Achieve Inc.

Saxon is a trademark of Harcourt Achieve Inc.

Printed in the United States of America
ISBN: 1-56577-960-6

4 5 6 7 8  446  12 11 10 09 08 07 06

## Phonetic Concepts Practiced

ch (chant, rich)
ee (sheep)

## Nondecodable Sight Words Introduced

where

ISBN 1-56577-960-6

Grade K, Decodable Reader 14
First used in Lesson 131